I0488855

Duckweed Ethanol:

Duckweed Biomass Grown from Organic Wastes to Replace Corn for US and International Ethanol Biofuel Production

by Christopher Kinkaid

Published by Solardyne, LLC
Portland, Oregon

ISBN-13: 978-1500485764
ISBN-10: 1500485764

Table of Contents

Preface

The smallest flowering plant, on Earth, is one of the most powerful, and widespread: duckweed. Usually, considered a nuisance, duckweed, upon close examination, is an impressive crop, in photosynthetic value.

Ethanol, an industry dominated by the Corn Industry (King Corn), faces many challenges, including large water draws, rising fertilizer costs, large diesel fuel bills, and unintended impacts on Food markets. Corn, as a choice for ethanol production, pits food, versus fuel, for agricultural resources, increasing stresses between fundamental markets.

An ideal source of biomass, for ethanol production, would not be a food crop, rather, a waste-crop.

King Corn, dominates current domestic ethanol production markets, worth billions, each year. Supported with Federal Farm Subsidies, worth billions of dollars annually, the corn industry dictates the US ethanol markets, using Corn as the principle feedstock crop.

At first glance, Corn, is an odd choice for ethanol production. Corn, began as a wild seed crop, domesticated by ancient man. Before the modern age, thousands of years of selective breeding, produced a Corn rich in proteins, and high in nutritional value.

Modern Corn, has been engineered to go "the other direction," and reduce Corn's Protein, and increase Corn's Starch (Carbohydrate) production. The "Starch" in corn, is used for Ethanol production, and other by-products, such as Corn Syrup, and Distillers Dried Grains and Solubles (DDGS).

Duckweed, is a choice for bulk biomass, which offers significant advantages over corn. Duckweed advantages include, lower energy costs, lower water resources, lower fertilizer costs, doesn't require valuable farmland, doesn't compete in Food markets, has higher Starch yield, per acre. Duckweed, in a controlled environment, can be grown, year round, and in diverse locations.

Corn, as a bulk source of Starch production, competes with Food markets, drinks thousands of gallons of water, per gallon Ethanol produced, requires large diesel fuel bills for growing, and harvesting, requires large amounts of fertilizers, and lower's the nutritional value of Corn on purpose, to produce more starch, reducing protein production, and nutritional value.

The market opportunities for Duckweed, displacing Corn are enormous. Emerging as the Second Generation biomass crop, Duckweed offers the best path, for tapping into the Ethanol production market, worth Billions of dollars.

About the Book

This Book is written to describe the next generation feedstock for the Ethanol industry. This Book presents a case to replace traditional corn crops, with Duckweed Biomass production from waste streams. The advantages, to Duckweed, over Corn, are numerous, and powerful.

Chapter One describes the big picture in Ethanol production. There are many pathways to Ethanol, from "Sugar," and "Starchy," plants. Corn, grown for starch, is the feedstock for present Ethanol production.

Chapter Two explores the Duckweeds, and how they can grow from nutrients, as well as discussion of Duckweed composition in amino acids, proteins, lipids (oils), and Carbohydrates (Starches), suitable for Ethanol production.

Chapter Three examines Ethanol in terms of input needs. Water consumption, fertilizer, mechanized planting and harvesting techniques. Plants grown for starch.

Chapter Four looks at growing, and using corn as a feedstock for Ethanol. Issues, of water needs, genetic engineering, fertilizing, and harvesting techniques. Competition with food markets.

Chapter Five describes Ethanol production from Duckweeds. Growth cycles, nutrient requirements, harvesting, nutrient manipulation.

Chapter Six looks at Duckweed economics. Fertilizers, water use, distributed growth, new nutrient streams open a competitive environment for the development of Duckweed Ethanol.

Chapter Seven examines the value of Organic Waste streams. Organic wastes, are valuable sources of Phosphorous, Nitrogen, Potassium, trace-elements, and micro-nutrients. Reduced to an in-organic form, through anaerobic digestion, or natural bacterial decomposition, organic wastes offer a vast resource for energy production.

Chapter Eight examines the global liquid fuels market. Gasoline, Aviation fuel, Diesel fuel for construction, farming, trucking, and general transportation uses, liquid fuels command a billions per day market.

Chapter Nine explores nutraceutical markets for Duckweed proteins, and lipids. Specialty feed markets, and major markets such as fish, foul, and animal feed markets.

About the Author

Christopher Kinkaid

Christopher (Toby) Kinkaid, originally from Portland, Oregon, is the founder of **Solardyne.com**, **SolarQuote.com**, and **AlgaeToday.com**, and has worked in clean energy technology for over three decades.

Kinkaid, is the inventor of the "**Helyx**" Vertical Axis Wind Generator, the "**Mariposa**" Non-imaging solar concentrator PV module (continuous operation at Sandia National Laboratory since 1994), the **Solar Demultiplexer** optical solar concentrating lens (Dr. James/Sandia National Laboratory 1991), and the inventor of the original "Solar Power Pack" (Mother Earth News, "Littlest Utility" June/July, 2001).

Kinkaid, has lectured on clean energy technology around the world including "APEC", Bangkok, Thailand, 2003, "Energy Solutions World", Tokyo, Japan, 2003, The International Biomass Conference (IBC), 2010, Minneapolis, MN, and the Algal Biomass Organization (ABO) Conference, 2010, Phoenix, AZ.

Christopher (Toby) Kinkaid, has appeared in interviews on KOIN TV, KGW TV, and "Sustainable Today" produced in Oregon, and has served on the board of directors for the National Hydrogen Association, in Washington D.C., 1993, the Japan Satellite Communications Company (JCNET), Fukuoka, Japan, 1994-95, and Algaedyne Corporation, St. Paul, MN, 2010-2013.

Kinkaid, presently serves as CEO of Solardyne, LLC in Portland, Oregon, where he continues his work in Solar, Wind, and Biomass Technology applications, research, and development.

Introduction

Ethanol, is an alcohol distilled from Sugars. Yeast, (Saccharomyces cerevisiae), as with all fermentation, is used to feed on Sugars, producing CO_2 gas, and ethanol. Ethanol, can be made from many different plants, as long as you end up with simple sugars. If you don't have sugar-rich plants, such as Sugar Cane, or Sugar Beets, you can use "secondary" plants as a feedstock, as long as they're "starchy."

The US Ethanol industry, largely based on Corn, uses corn, as an indirect source of sugar, by first growing Starch.

Corn, not a particularly sweet plant, is used by present industrial interests by "engineering" corn to be more "starchy," actually lowering the nutritional value by reducing protein production.

Genetic engineering, over the last-half century, have intentionally, converted strains of corn to stop converting proteins, and make starch instead. Engineering corn, to lose protein, (largely valuable molecules), and replace them with "starch," a much less valuable molecule, seems odd, at first glance.

Why, would industry want to convert Corn into Ethanol?

Starch, it turns out, can be turned into sugar, with an intermediate step (Saccharification). Starchy plants, rich in Carbohydrates, such as potatoes, grains, and roots, can be converted, in part, into sugars using a special Enzyme (alpha-Amylase). This special Enzyme, feeds on the starch, and produces CO_2 gas, and Sugars. Starch, turned into sugars, with this Enzyme, are then fermented with Yeast, in the age old process, and Ethanol is produced. Corn, goes from Starch, to Sugar. Then, Sugar, to Alcohol.

Is there another path to Ethanol?

Since, we know we can use a "food" crop for ethanol production (Corn), can we use a "Non-food" crop instead?

Yes. It turns out, you can use any plant you wish, if it meets your standards of "starchiness" for these purposes. The question begs: then what are the best plants to use?

Ethanol from Duckweed offers a pathway which is organically based. Duckweeds ability to uptake nutrients, from "polluted" sources, with less water use, less machinery required for harvest and processing, and grown from distributed sources such as animal feed lot operations, offer significant advantages for growing Duckweeds, as biomass for ethanol production.

Chapter One - Biofuels Versus Fossil Fuels

The modern industrial age, from the First Industrial Revolution through today, powers on Fossil-fuels.

Fossil-fuels, dictate a world of "Haves," and "Have-nots," from the simple fact these materials are not evenly distributed in the world.

The problem with burning Fossil-fuels, for base load electricity, and transportation energy, on industrial scale, are toxins, and unequal access for citizens worldwide.

Our civilization, from the 1st Industrial Revolution, with coal fired steam engines, through today, has been powered by Heat-Engines, and their ever increasing hunger, or thirst, for fuel to burn.

Combustion, by definition, at any temperature, other than the extreme, cause partial combustion products, producing a concentrated emission of toxic vapors. For every 1 Kilogram of Carbon burned, at least 2.2 Kg of CO_2 gas is formed. For every ton of carbon fuel burned, such as coal, more than 2.2 tons of CO_2 gas is formed, and emitted into the biosphere.

The combustion of fossil-fuel Carbon, binds oxygen in the atmosphere, with Carbon producing vast amounts of CO_2, NO_x, SO_x, particulates, and other partially combusted hydrocarbons, acids, and even radiation (from concentrated coal ash).

Worldwide, millions of tons, Per Day, of Carbon-based fuels are burned, producing a wide distribution of critical entry points into the environment. En mass, these emitters contribute million of tons of toxic emissions, daily, to support our current civilization.

There is a great risk in over stressing ecosystems, and species, too quickly. A famous Biologist once said"Species, when stressed, either adapt, or die."

Our civilization is at a cross-roads.

Not all Carbon is the same

Ancient carbon resources, Coal, Oil, Natural Gas, all took millions of years to form. Yet, our civilization burns them for a moment's power.

There are two "kinds" of Carbon on Earth. Ancient Carbon, and present Carbon.

Ancient Carbon, was formed over millions of years from Biomass produced from ancient solar powered photosynthesis. Through Deep Time, these organic materials, buried, and subjected to intense pressures, and temperatures, produce our current Fossil-fuels.

Fossil-fuels, were formed in a "Goldilocks" situation. Pressure-cook organic material too long, you get coal. Pressure-cook organics too short a time, and you get Shale. Pressure cook organics, just right, and you get petroleum.

Aside, from the geopolitical stresses, of one country having more "access" to fossil-fuels, over another, the largest physical danger is toxicity.

Higher life forms (more complexity), such as sea mammals, deep sea fish, and humans, all exhibit the characteristic of "bioaccumulation." Perhaps, the scariest word in the human language, bioaccumulation, is the tendency of higher life forms to accumulate toxins.

Basing a civilization on fossil-fuels, presents a scenario of toxicity, saturation, and collapse.

Toxicity is unforgiving. The greatest danger presented by continued Fossil-fuel dependence, on a global scale, is toxicity-induced biological collapse.

Ethanol, is an attempt to replace liquid transportation fuels, using crops as the feedstock.

"Present" Carbon, is Carbon used in the "hydrosphere," cycles involving the Atmosphere, and the Oceans. Carbon, "fixed" into molecules of life, by photosynthesis, draw Carbon from CO_2 in the atmosphere. Plants, when consumed, or burned, release this Carbon back into the atmosphere to be recycled again.

Carbon-Neutral, the design goal, is to source Carbon from the atmosphere (growing plants), by "fixing" carbon (Carbon Fixation). Sourcing Carbon, from the atmosphere, as plants do, allows the CO_2 produced, when consumed, or burned, to be returned to the Atmosphere. This is Carbon-Neutral. Note: There is a constant exchange between the Atmosphere, and Oceans.

Ethanol, can be produced, from any Sugar, or Starch (with another step), and can use many plants, as a biomass feedstock. Ethanol, produced from Duckweeds, offer a real improvement in biomass productivity, and usefulness, in mitigating organic wastes, and producing Ethanol fuels, feeds, and fertilizers.

Ethanol production, using Corn, as the principle source of "Starch" offers large opportunities for a new agricultural technology to move into the Ethanol production market. Duckweeds, as the chapters below will describe, offer advantages over Corn, and other biomass crops, for the production of organic Fuels, Feeds, and Fertilizers.

Chapter Two - Duckweed the Plant

The Plant Kingdom drives life on Earth, as the "engine" of Oxygen, and Nutrition. Without plants, we don't eat, or breath. The base of the food web, on Earth, and in the Oceans, are the autotrophs. The Plant Kingdom, on Earth, is the primary engine producing Oxygen, and Base Nutrients, on which, all life on Earth is sustained.

Life on Earth, is driven by oxygenic photosynthesis. Converting, in-organic mineral salts, CO_2, water, and selected wavelengths in sunlight, photosynthetic organisms do the incredible: synthesis sugars, carbohydrates, and proteins, from in-organic materials.

Duckweed, the simplest, and smallest flowering plant, grows all over the world in fresh, brackish, and salt waterways. Preferring the tropics, and sub-tropical regions, Duckweed also flourish at higher latitudes, common during Summer months as far North as Canada. Duckweeds, have evolved to thrive in "eutrophic" water conditions, and present almost a "super-plant" ability to uptake nutrients, from brackish water sources.

Autotrophs, and Heterotrophs.

Essentially, there are two kinds of "life" on Earth (over 99.99% by mass): autotrophs, and heterotrophs.

Heterotrophs, like us, require eating, to gain essential nutrition. Without eating, carbon based molecules for nutrition, Heterotrophs, (including all insects, birds, fish, reptiles, amphibians, and mammals (humans) can't survive.

Autotrophs, (the Plant Kingdom), however, do the remarkable, by "manufacturing" their own "food" directly from selected solar energy wavelengths, mineral salts, trace-elements, CO_2, and water.

Photosynthesis, using the primary pigment, Chlorophyll, power's the Earths Food Web, and is the most important process sustaining all forms of life. Oxygenic photosynthesis, oxidizing water, and reducing CO_2 into glucose molecules, releasing oxygen in the process, fuels our natural world.

From the Plant Kingdom (autotrophs), all basic molecules of life, including amino-acids, proteins, lipids (oils), carbohydrates, and complex molecules like enzymes, anti-oxidants, are all created by the process of Photosynthesis. Duckweed, is an amazing embodiment of photosynthesis converting in-organic molecules, into organic compounds, with incredible speed.

Growing Duckweed

Duckweed, is the smallest flowering plant on Earth, and floats on water. The enormous value of Duckweed, is its ability to extract, and convert "mineral salts," and other inorganic materials, dissolved in water, through Photosynthesis, to produce Duckweed biomass. Doubling in mass, depending on conditions, in about 48 hours, Duckweed is an impressive life form.

Aquatic plants have been harvested for animal, and human consumption for thousands of years, and Duckweeds have a long history with people. A number of aquatic plants, offer superior nutritional value, and are (were) harvested throughout human history. Harvesting free floating plants such as water lettuce, (Pistia), Azolia, water hyacinth (Eichhcornia), and Duckweed (Lemna), has proven to be a great source of Vitamin A, trace elements, anti-oxidants, and amino-acids.

As most aquatic species, Duckweeds, are very sensitive to their growing environment, and can be "stressed," (selective growth protocols), to produce either larger amounts of Proteins, or, large amounts of Carbohydrate (starch), depending on growth conditions.

Duckweeds, in general, when dried and ground, is found to contain roughly 1/4 protein, 1/3 Carbohydrate (excellent for starch to change to sugar for fermentation), and almost 1/6 Lipids (Depending on Growing media used).

Lipids, rich in Omega III, and VI, oils are highly valued in the nutraceutical, animal, fish, poultry, and human nutrition markets, Duckweed is a great source of these important lipids. Duckweeds, produce essential amino-acids, highly prized in the nutraceutical markets.

Duckweeds, can be "influenced," by adjusting N:P:K ratios in the growth media, as well as levels of ammonia, which make Nitrogen available to Duckweed. Duckweed, for growing biomass, can be influenced to produce as much as 75% (Dry-weight Mass) Carbohydrate (Starches, and Sugars).

Duckweed, has an ideal chemical makeup for three major industries: feed, fuel, and fertilizer

Duckweed, grown in a controlled environment, produce a superior source of starches, with proper nutrition management, and makes the case for consideration as the 2nd generation feedstock for Ethanol production.

Duckweeds, are small fragile, free floating aquatic plants, and grow best in still, shallow water with elevated N:P:K levels. These characteristics, make Duckweeds, a viable commercial biomass crop for treating waste-water, and producing high starch content for processing into Ethanol fuel.

Duckweeds belong to five genera; Spirodela, Lemna, Landoltia, Wolfia, and Wolffiella. Of these genus, nearly 38 species, (taxa), are known worldwide.

All species of Duckweed, have somewhat flattened, oval "fronds" from 1 mm to nearly 9 mm in diameter, depending on species, and reproduce asexually, and sexually. Free floating, simple flowering plants, some species of Duckweed develop shallow root systems, which help the plant absorb nutrients, and some researchers report, helps stabilize the plant in the water column. The fonds of Spirodela, and Lemna, are flat, oval and resemble leafs. Plump, and gas filled, for buoyancy in the water column, duckweed grow best on the surface, or very near the surface of the water.

Duckweed species are often found to coexist with each other, and found all over the world in natural ponds, lakes, streams, including brackish, and saline pools.

Duckweed, has been successfully grown in water temperatures between 6-33 degrees C. Optimum temperatures, seem to be close to 30 degrees C. The water chemistry, or growth media used is vital from a pH standpoint. Duckweed, grows between pH 5-9, but grows best over a range between 6-7.5 pH. Optimum pH levels between 6.5 and 7.

Note: a challenge for growing Duckweed, in shallow ponds, is keeping the temperature constant, and in desired ranges. Great care, should be taken, in designing your cultivation method to keep all parameters in optimum ranges.

Duckweed, is a mineral, and heavy metal sponge. Used, as the 'canary in the mine', Duckweed samples, are often used as measure of water quality. Duckweeds, ability to absorb, and uptake heavy metals, Macro, and Micro-nutrients, is a double edged sword.

Duckweeds, uptake N, P, K, Zn, Sr, Co, Fe, Mn, Cr, Cd, Cu, Pb, Au, and Al. Analyze your source water for profiles.

Used in Water, and waste treatment facilities, duckweed used for fuels, can be used to absorb out of balance nutrients (N:P:K), out of water systems, providing a benefit (and income stream). Note: Duckweed, used for the nutraceutical markets, need to monitor source water for heavy metals.

As for most plants, Duckweed, needs many nutrients, and minerals to grow rapidly, and in proper balance. In nature, slowly decaying organic material, attacked by oxygenic bacteria in the water column, provide a steady stream of nutrients, and Duckweed, has evolved to maximize this advantage.

Growing Duckweed, commercially, requires key elements to be balanced in the growth media. For example, the following compiled concentrations, are found to be beneficial, as a high growth media can be mixed. Duckweeds, absorb, convert, and amass nutrients from the source water into biomass.

The following chart examines concentrations achieved with Duckweed:

Element	Growing Medium (mg/l)	Duckweed (mg/kg Dry Mass)
Nitrogen	0.72	59,000
Phosphorous	0.32	5-14,000
Potassium	99	39,100
Calcium	357	10,000
Magnesium	74	5,900
Sodium	247	3,240
Iron	99	2,390

Duckweed, is very beneficial for higher life forms, with a complete amino-acid profile. A sample specific composition of Duckweed (Amino-Acids) is described below (Dry-Weight Percentage):

Component	Lemna	Soybean
Crude Protein	27%	43%
Lysine	3.5%	6.5%
Histidine	1.6%	2.4%
Aspartate	-	-
Arginine	5.1%	7.2%

Component	Lemna	Soybean
Threonine	4.2%	3.8%
Serine	-	-
Valine	5.8%	4.5%
Methionine	1.5%	1.1%
Leucine	7.7%	7.6%
Isoleucine	4.2%	4.4%
Tryptophane	4.1%	3.5%

Duckweeds, are rich in amino-acids. This protein profile is valuable to fish, and poultry farmers, which need nutritional supplements in their feed. Fish diets, often lack vital amino-acids. Research, and long practices in growing Duckweed in Vietnam, for example, has found Duckweed, suitable for up to half of the feed requirements.

Duckweed, can be grown with a source of nutrients, water, and solar energy. Improving the output achieved in nature, careful cultivation of Duckweed, in a controlled environment, can achieve higher yields.

Duckweeds, can be "manipulated" in the growth phase, to maximize the production of particular elements. Growth protocols for increasing Proteins, or increasing Carbohydrates (starches), are dictated by your growing cycle, and how you "limit" certain nutrients.

Duckweeds, are a treasure engine of productivity. The widespread commercialization of Duckweeds, poises Duckweeds to offer a profitable choice in Ethanol biomass feedstocks.

Chapter Three: Ethanol the Big Picture

Ethanol, is an alcohol produced from distillation. Going back, as far as we considered ourselves human, distillation is one of the oldest human arts. Physical evidence for distillation dates to the Mesolithic, and probably extends back to the Paleolithic.

Ethanol, (ethyl-alcohol) is fermented from Sugars, which can be used directly, or derived from Starches.

In modern times, distillation, used for the brewing, and liquor industries, produce billions of gallons of spirits, and it's this basic process, which is used in the Ethanol world.

Corn production, has dominated the ethanol industry, securing economic subsidy, as the principle feedstock for Ethanol production in the US.

Corn, grown for Ethanol production (usually Yellow Corn), is grown, principally, for Starch. In the Wet-mill process, a pretreatment separates the Starch components from proteins, and lipids. After fermentation, (the proteins are largely un affected during fermentation), the Ethanol is distilled and separated. The remaining "Stillage" produces a condensed Corn Syrup (Liquids), and Dried Distilled Grains, and Solubles (DDGS).

Ethanol, can be made from any "sugary," or "starchy" plant. And, the choice of plant, used for Ethanol production, can be localized to take advantage of local micro-climates. Sugar beets, and Sugar Cane, grow well in tropical regions. Potatoes, and Root plants grow well in moderate, and cooler climates. However, in the US, Corn has been the dominant "feedstock" for growing Starch.

The aquatic species, Duckweed, in the popular vernacular, grow world wide from tropics, sub-tropics, and into higher latitudes during Summer months. Duckweed, being ubiquitous worldwide, is an excellent choice for commercialization. A controlled environment, appropriate for commercial growing, offers increased yield to offset the capital investment in hardware. Controlled environments dramatically increase Starch production per acre cultivated, and would allow production throughout the entire year.

Corn, can only be planted once a year. Corn, with an average growth cycle of 120 days, allows only one Crop to be harvested per year. Prime farmland, can only be productive 1/3 of the year.

Duckweed production, using controlled environments, on marginal land, would produce a year-round production, and produce many times more Starch yields, than single mono-crops of corn.

Ethanol, a distilled process, can be sourced with a wide variety of plants.

The replacement feedstock, often called the Second Generation fuel, will not be based on Corn, but on some crop which displaces Corn.

To displace Corn, the Second Generation Feedstock will need to bring economic leverage to bear. Displacing corn, as the principle feedstock, will require demonstration of life-cycle, economics.

Ethanol production, is based on the fermentation of sugars. Sugars, can be derived from Starches, by addition alpha-amylase enzyme.

Corn, has become the dominate feedstock in the US Ethanol industry.

This eBook is written, to make the case that primary Ethanol production, will begin to shift from corn, to Duckweeds, as the primary feedstock crop.

Competitive Advantages to Duckweed over Corn

Corn, is a food crop. Duckweed, is a non-food crop. Food crops, used for fuel production, always put unwanted pressure on Food markets. Increasing demand, increases prices.

Recent experience with rapidly rising "Yellow" corn prices, had dramatic markets for corn tortillas in Mexico. Duckweed, being a non-food crop, would have no impact on Food markets.

Rising demand for corn, used for food, will always provide a market for corn producers.

Corn, requires thousands of Gallons of Irrigation Water per Gallon of Ethanol produced. Duckweed, requires, depending on cultivation techniques, and grey-water recycling, about 100 times less water, per Gallon of Ethanol produced. Duckweeds, as such, do not require large irrigation, and, in controlled environments, are not subject to the massive "evaporative" losses, faced by large irrigation of corn.

Corn, used for Ethanol production, typically, require fertilizer. Fertilizers, usually derived from Fossil-fuels, present increasing costs to corn farmers. Also, fossil-fuel fertilizers, increase the "Carbon Footprint" of traditional corn based Ethanol.

Duckweeds, derive their bulk nutrition, from effluent run-off from a host of industries which produce organic waste. Dairy, Hog, Poultry, and Fish farms produce enormous volumes of organic wastes, well suited for Duckweed "fertilizer." Duckweeds incredible ability to uptake nutrients, from water streams, produce an "Income" stream, as bioremediation is an acknowledged industry.

Corn, requires many "passes" over the land by large mechanized machines, for tilling, seeding, fertilizing, and harvesting corn crops. Traditional corn farming, presents farmers with large "Diesel" fuel costs, and further increases the Carbon Footprint of corn-based Ethanol production.

Duckweeds, do not require, large, or powerful, equipment. Controlled environment, often gravity fed, provide much lower energy demands, in cultivation, and harvest activities. Traditional diesel fuel costs, present another variability to corn production. Reducing fuel requirements for starch production, gives Duckweeds another strong economic advantage, in competition with corn.

Corn, having matured over the last two decades, cannot escape the fundamental market forces. Subsidies, will continue to prop-up corn producers. However, the systemic advantages of Duckweed, will soon find a beachhead.

Large AG industries, such as Dairies, Cheese factories, Hog, Poultry, and Fish farms, to mention a few, have enormous values in their waste-streams, under, or non-utilized. The production of Duckweed-based biomass, from these waste streams, presents an enormous market opportunity, for these AG operations.

Chapter Four: Ethanol from Corn

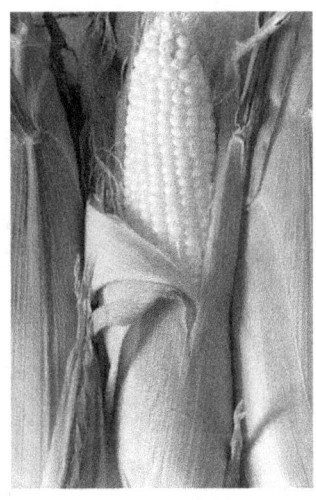

Ethanol, from Corn, has a limited future. Second Generation feedstock crops, coming online, present a potential threat to Corn's dominance. At present, Duckweeds, offer a clear crop choice which would replace the Corn Industry. Corn, as a feedstock for Ethanol, presents several issues for sustainable growth, and therein lys the potential for a "disruptive agricultural technology" which presents more added value, with lower costs, to come in and change the fundamental feedstock of US Ethanol production: corn.

The opportunity for Duckweed to move in is overwhelming, and will prove to be resisted, then embraced, as market share develops, and systemic advantage to Duckweed come to bear.

Corn, as a feedstock for distillation, presents several drawbacks.

Water Resources

Corn, grown for Starches, drinks enormous volumes of water, in the farm belt. USDA estimates that 3,000 gallons of water is required for each 1 Gallon of Ethanol produced, from watering crops, if those crops are Corn. Water, also consumed in the Ethanol distillation process, with some grey-water recycling, require additional water inputs. Water consumption, in distillation varies, but industry averages range around 2.5 Gallons of water per Gallon Ethanol produced, in fermentation, and distillation.

Water resources, used for crop watering, is the major consumer of water in the Corn crop scheme. The production of 1 billion gallons of Ethanol, from

Corn, requires 3 Trillion Gallons of Water for crop consumption.

Genetic Manipulation of Corn to Constrain Protein

Genetic Modified Organisms (GMO) dominate the strain selection process of Corn being planted throughout the Midwest. Monocultures, in general, and specifically for Corn growth, destroy biodiversity raising risks of invasive species, and pests.

Proteins, being reduced in Modern genetic engineering of Corn, reduces the nutritional value systemically, causing a decrease in net value-added by the Corn Farmer.

Water pumping is energy intensive. Any Life Cycle analysis of Corn based Ethanol, must also include the energy costs, for irrigation water pumping, tractor tilling, fertilizing, seeding, and harvesting.

Fertilizers

Modern Corn growers, when growing for Starch production, typically use fertilizers, increasing costs. Fertilizers, derived from Fossil-Fuels, increases the cost of Corn production, and increase the local pollution impacts of run-off into natural waterways.

The need to fertilize Corn-based monocultures, or mono-crops, depletes the soil of vital nutrients, and once Fossil-fuel based fertilizers are used, it produces a vicious cycle of continued need.

Diesel Fuel Consumption

Growing Corn, requires Farm Equipment. Tractors, till the land, spread the fertilizer, spread the seed, harvest, and till the ground again. It takes a hand full of passes, over each acre of cultivated Corn, by Farm Equipment, heavy equipment, that drink Diesel fuel.

Ask Farmers, what their Diesel Fuel Costs are each year, if you want to see a Man turn into a Ghost.

Food Market Pressures

Yellow corn is engineered to be starchy for the Ethanol, and Corn Syrup markets. Unfortunately, using food crops for fuel, causes prices for Food Markets to spike upward, as more demand pressure is applied by Ethanol consumers.

Increased pressure on Food Commodity markets, are an unfortunate impact of choosing Corn as a fuel feedstock.

Most Ethanol, distilled in the US, is derived from Corn.

Corn, can be processed, into Ethanol, and other products using either Wet, or Dry Milling methods.

In the Dry-milled approach, the entire Corn kernel is finely ground into " Corn meal." Water, is added to the meal to form a slurry, or "Mash."

Special enzymes, (Amylase), are added to the Mash, to convert the Starches (Carbohydrate Polysaccharides), into simple Sugars (usually dextrose).

The Amylase enzyme, is used in the human body, in both saliva, and in the pancreas. When you chew a starchy food, such as potatoes, you may taste a light sweetness at the end of chewing. The Amylase, in your saliva, has already begun to break down the starches into simple sugars, which you can detect with your taste buds.

In Wet-milling, Corn kernels, are soaked in a dilute solution of Sulphuric Acid to break up the biomass into components. Once, soaked, and ground, the Corn kernels break down into constituent molecules, composed of crude proteins, starches, and lipids.

Wet milling is used to break down the Corn into components, and separate those components for different markets. Starches, are removed for Ethanol processing, while other products, such as Corn Oil, and DDGS are removed, and sold, into the Animal Feed markets.

Valuable Farmland

Corn, is grown on the most valuable farming land in the US. Capable of growing many diverse crops, farmers who grow corn, grown this mono-crop for starch production. Duckweeds, can be grown on any type of land, being an aquatic species. Marginal land, is most adequate, for Duckweed cultivation, and presents another major advantage compared with growing corn.

Chapter Five - Ethanol from Duckweed

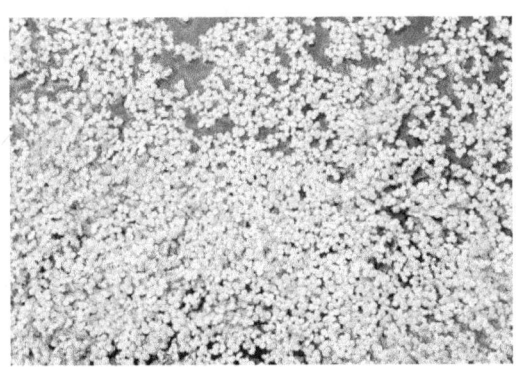

Ethanol, an alcohol, can be distilled from any "sugary," plant source.

If you don't have a "sugary" plant, you can start with a "Starchy" plant, and convert the mash to Sugar, using special enzymes (alpha-Amylase). The objective, for any fermented product, is to start with a Sugar biomass, even if you started with "Starchy plants" to get to the sugar phase. Once you have sugar, you can ferment.

Ethanol, from Duckweed, will follow a similar chemical path, as used for Corn.

Duckweed can be grown from waste-water streams (no need for fertilizing), and grown to produce Starch components for harvest. Duckweed, yields of dry-mass, per acre, per growing year vary with location, and management. Typical Duckweed yields, average about 15 tons dry-mass per acre, per year.

Duckweed, can be processed into its three major components in a process involving mashing, and grinding duckweed biomass, and dewatering. Once, Duckweed has been ground, and dewatered, a Centrifuge, can be used to separate different materials present in the mash, for harvesting.

The Centrifuge, accelerates the mash, inside the rotor, to very high speeds. Centripetal forces, inside the rotor, cause the different densities, of materials, to separate into layers, with the outside layer the most dense. Proteins, are generally heavier than Carbohydrates. Carbohydrates, are generally heavier than Lipids (Oils).

Ethanol, can be produced tapping into the approximate 75% dry mass of Duckweed as Carbohydrates (Starches) suitable for enzyme reduction into Sugars. Note: proper growing media manipulation is required to reach these high starch contents.

Sugars, enzymatically derived from Duckweed Carbohydrates, can be distilled into Ethanol, using the age old process of Yeast fermentation (Saccharomyces cerevisiae).

The remaining Proteins, and Lipids, can be separated, and sold to the animal, poultry, and fish feed markets. Dry weight value of proteins, varies with the market, but can be approximated at $1,200 per ton (wholesale), however, this value fluctuates with market rates.

Duckweeds, are aquatic species, and can be grown on marginal land. Duckweeds, uptake nutrients from waste water streams. Duckweeds, can be grown manipulated to increase Proteins, or Carbohydrates for Ethanol production.

Duckweeds, require hundreds of gallons of water, per gallon of Ethanol produced, than corn irrigation.

Duckweeds, under growth protocols to increase starch production, can produce many times more starch, per acre, than Corn.

These advantages, add up to a competitive scenario for Duckweeds, in competition with Corn crops for Ethanol.

Researchers, report different rates of conversion of biomass for Ethanol, based on the specifics of their nutrient mix, growing cycle, limiting of specific nutrients, at strategic times, and frequency of harvest.

Ethanol production, from Duckweed biomass, has reached 0.5 grams of Ethanol, per gram of Duckweed starch. Duckweed starches, are most suitable for Ethanol production, and represent a major leap in sustainability.

Chapter Six - Duckweed Ethanol Economics

Ethanol, produced with Duckweed biomass, is dictated by the same economics, all Ethanol producers face, with some exceptions, and advantages. Supply, and Demand, is no small matter in the Duckweed Ethanol economic picture, and offers several advantages to Duckweed biomass, in competition with Corn.

Duckweed, enjoys several strategic advantages, over Corn, as a feedstock, including diversity of source, mitigation of waste streams, high starch content per acre, CO_2 Carbon tax credits, Carbon production tax credits., Protein byproduct suitable for feed markets, Ethanol subsidies, and of course, the value of the fuel on the open market.

Duckweed can be grown, in controlled environments, producing six times the Starch density of Corn. Duckweed, can be sourced,

(Grown) anywhere in the US there are organic wastes, with properly engineered growing apparatus.

Duckweed, requires no traditional nutrients from buying "fertilizer". Only nutrient "balancing" is required for commercial Duckweed cultivation. Duckweed, grows in water, but very shallow water. Duckweed, cultivated in controlled environments, can recycle almost 90% of grey water for further cultivation, greatly reducing the water resources required.

Duckweed biomass, can be produced with several income-streams, not enjoyed by Corn. Income from mitigating waste, further income from Carbon tax, and production credits, due to displaced fossil fuel consumption.

Duckweed, though edible, is not considered a "Food" crop, as such, would have no unintended impacts on Food Commodity markets.

Corn, in the US, grows best in the Midwest region. Duckweed, in the presence of organic wastes from Animal Farms, and CAFO operations, can be grown in all 50 states. Duckweed, grown in all States gives a widespread distributed fuels supply base. Duckweed, grown in a controlled environment, can be cultivated year-round. Further, increasing the net yield from each acre used to grow Duckweed.

Corn, has only one crop per year, and takes over 120 days to reach maturity. Duckweed, can have multiple crops, per year, and in a controlled environment for Northern climates, can produce Duckweed biomass year round.

Duckweed, can be grown on marginal land, not suitable for traditional agriculture, greatly increasing the land available for cultivation.

Energy security, would be enhanced with a diversified energy base. Biofuels, and particularly Ethanol production, from Duckweed biomass offers a diverse, and active job base for all states. Diversity in the Ethanol production market, creates jobs in all 50 states, and encourages consumption of locally produced fuel, by local markets.

Consumers, using locally produced Ethanol fuels, would greatly reduce foreign energy supply price manipulation, and stimulate local economies. Localized Ethanol production, using Duckweed grown with nutrients from Organic Waste Streams

Corn, as an Ethanol feedstock has many costs, and variable costs to navigate. Duckweed, has costs, but offers income-streams unavailable to corn production. Mitigating organic wastes, and the associated carbon reduction has value. Income, can be generated by "treating" waste water with Duckweed growing ponds (or environments).

Using the nutrients, removed from the waste water, duckweed synthesis photosynthetic products of great value, namely, Proteins, Carbohydrates, and Lipids (oils).

Duckweed, is a choice for bulk biomass which offers significant advantages over corn. Duckweed advantages include, lower energy costs, lower water resources, lower fertilizer costs, doesn't compete in Food markets, has higher Starch yield, per acre, can be grown, year round, and in diverse locations.

Corn, as a bulk source of Starch production, competes with Food markets, drinks thousands of gallons of water, per gallon Ethanol produced, requires large diesel fuel bills for growing, and harvesting, requires large amounts of fertilizers, and lower's the nutritional value of Corn on purpose, to produce more starch.

Corn, is the current leader in biomass production for Ethanol, but can only be through large subsidies of public funds. Duckweed, with many value streams, offers a competitive advantage compared to Corn, and with proper growth protocols, will begin to gain market share.

Duckweeds, offer a path from waste-streams, into revenue streams. Adding value, by treating waste water, removing Phosphorous, Fixed Nitrogen, and Potassium, including trace-elements, mineral salts, and other important nutrients, Duckweeds, produce a Crop which is easy to harvest, high in valuable proteins, and rich in suitable starches for Ethanol production.

The Economics of Duckweed commercialization will be based on multiple value streams. Duckweeds, rich in Proteins, Carbohydrates, and Lipids convert wastes, generating income, into Fuels, Feeds, and Fertilizers.

Integrating Anaerobic Digesters, as a pre-treatment for organic wastes, produces an organic cycle which converts organic waste streams into Electricity, Ethanol, Proteins, and Lipids for Feed markets.

Effluent, from the digester is used as an excellent fertilizer, rich in in-organic compounds, readily available for plants.

The economic success of Duckweeds, will depend on the many "added values" achieved with Duckweeds. Listed above, from fertilizer, water use, and starch density grown, Duckweeds offer a superior feedstock for Ethanol production.

Duckweeds, grown from animal wastes, present the greatest opportunity for diversifying the Ethanol fuel base, in the US, as well as in international markets, with increased local supply. Animals, (fish, fowl, farm animals), present a goldmine in nutrients, easily converted by Duckweeds into valuable molecules, including Amino-acids, proteins, carbohydrates, and Lipids (oils).

Chapter Seven - Ethanol Fuel from Waste Streams

One of the best kept "industrial" secrets, is the enormous value present in Organic Wastes. Ethanol, produced from Duckweed Biomass, uses a similar chemical path to Corn, with several advantages. Duckweed, grows on nutrients present in Organic Wastes, and doesn't need to be purchased by the Duckweed grower. In fact, the Duckweed grower can be paid for removing Organic Phosphorous, Nitrogen, and Potassium compounds from waste water.

Duckweed, converts Waste-streams, into Revenue-Streams.

Organic Waste Streams, are a goldmine in valuable nutrients. Phosphorus, is highly valued, as a limited, and vital chemical for Fertilizing Plants. Nitrogen, fixed in compounds, is another valuable group of molecules present in organic waste streams. Worth thousands of dollars per ton, nutrients are discharged, and wasted, in many common practices, such as "Broadcasting" raw manure in the fields. Or, allowing Animal Wastes to enter natural waterways causing septic, and toxic conditions leading to local ecological collapse.

Duckweed, eat inorganic wastes, so to speak. Ponds, which are eutrophic, and hypoxic, are ideal conditions for Duckweed. Examine your nearest Pond in the Summer. Find a quiet, shaded area, and you're bound to see Duckweed.

Conditions, which would choke other plants, are ideal for the Duckweed plants, evolved to thrive in eutrophic conditions.

Waterways, in the US are becoming choked with a constant stream of untreated organic wastes. There are many famous examples, the Chesapeake Bay, stands out. The poultry, and swine growing operations, along the Chesapeake Bay, produce enormous run-off volumes of organic wastes.

Untreated organic wastes, are rich in N:P:K. Nitrogen, Phosphorous, and Potassium, as well as a host of other molecules, ranging from hormones, to enzymes, all spilling into the natural waterways, causing hypoxia. Eutrophic water ways, those with an "imbalance" of nutrients, are highly disruptive to native species.

Duckweed, grown from these waste-streams, is an excellent way to take a "problem," and turn it into a resource "solution." Duckweed, evolved so successfully, because decaying organic material, plant, and animal, always present in the waterways, provided excellent nutrients, for Duckweed growth.

Technically, decaying organic material is decomposed by primary decomposers, mostly bacteria. Bacteria, render organic waste, into in-organic forms, ideal for uptake by plants.

Duckweed, has evolved, a symbiotic relationship with aqueous bacteria, to thrive taking advantage of oxygenic bacteria in the water column, producing inorganic nutrient compounds, and oxygen.

Note: lakes, and ponds, become hypoxic (without oxygen) because Bacteria has consumed most available dissolved oxygen in the water. Depriving native species of dissolved oxygen, who depend on oxygen to breath, and respirate, causes species collapse. Algae blooms were thought to cause the hypoxia, but its the bacteria which feed on the Algal bloom, once the nutrients have been depleted, which cause the hypoxia.

Organic waste-streams, are available nearly everywhere. Breweries, Food processors, Animal,

Poultry, Fish operations. Even restaurants, supermarkets, and other commercial sources of organic material can be processed to grow Duckweed biomass.

Duckweed, requires a primary decomposer to break down the organic molecules, into in-organic form, ideal for plants to absorb. The easy way to do this is throw some organic material into some water. As the bacteria breakdown the organic material, into in-organic compounds, duckweed can thrive.

There are many paths, to convert organic wastes, into high value nutrients.

Anaerobic digesters, are an ideal method of utilizing organic waste-streams. Dairy operations, for example, have excellent success converting urea, and manure into a slurry which can be digested.

Anaerobic digesters, come in several forms. A simple, high performance design is called "Plug-Flow" type.

In an Anaerobic tank, or bladder (without exposure to air), urea and manure (after pre-settling in a settling pond) is pumped into the main digester Vessel. Internal augurs, move the material slowly through the tank, usually over 30-40 days.

During that time, organic material has been attacked by "anaerobic" bacteria, which eat the organic molecules belching Methane gas, and some

CO2. Methane, (Biogas), is then separated, and used as fuel.

The important advantage, is Anaerobic digesters convert Raw Organic Wastes, not only into Methane, which can be used as a Carbon-Neutral fuel, but also produces the high value "effluent" material left over. Digester effluent is rich in In-organic forms, which allow plants to absorb, and "uptake" the nutrients readily.

Plants, can absorb, and use, nutrients in in-organic form. Plants, through photosynthesis, "fix" carbon, by building complex organic molecules (starting with glucose), by Oxidizing water, and Reducing CO2, and use trace elements throughout the Light-Dependent, and Light-Independent sides of Photosynthesis. Trace elements, vitamin B, and micro-nutrients, are required to grow all plants, including Duckweed, so nutrient management is key.

The Calvin-Benson cycle, describes the physical process of photosynthesis producing biomass through carbon fixing, utilized, of course, by Duckweed.

The Business of Ethanol

Traditional corn feedstocks, require prime agricultural land, enormous water resources, fossil-fuel based fertilizers, fossil-fuel based diesel fuels for farm equipment, and competes with traditional "Food" markets.

Duckweed , based feedstock, uses "marginal" land, modest water resources, uses fertilizers from organic waste, doesn't require large farm equipment for growing, and doesn't compete with "Food" markets.

Duckweed, turns costs, into income streams, leading to systemic advantages in Ethanol production.

Chapter Eight - International Fuel Market at $120 Billion per Day

Transportation markets for Liquid Fuels, are worth billions of dollars per Day in the US. The international market, for transportation fuels, including Gasoline, Aviation, and Diesel fuels, is over $120 Billion per day. That's $120,000 Million, Every 24 hours.

The US Energy Information Office estimates 120 Million barrels per day, oil consumption, worldwide. Valuing a barrel of oil at $100, the liquid fuels market is worth $120 Billion per day.

This enormous market power, (Cash-flow), dominates governments, and international relations, and has played a leading role in modern history through two world wars, the Cold War, and into recent adventurism in the Middle East since 1972.

Coal, Oil, and Natural Gas commands Billions per Day, and shapes the long term energy consumption for Planet Earth. Toxicity, Bioaccumulation, and Economic Inequality all point to a replacement of the Fossil Fuel industries. However, looking towards the existing fossil-fuel industry, to be the agent of change, is short-sighted.

Fossil-fuel interests, are well aware, of the enormous economic power of fossil-fuel reserves. However, recent awareness of the toxicity, climate impacts, and economic disparity, has driven some interests to re-examine the fossil fuel world.

Duckweed, is a natural powerhouse of photosynthesis. All chemicals, derived from fossil-fuels, can be derived "organically" through photosynthesis. After all, the majority of fossil-fuels, are ancient biomass, from photosynthesis.

Duckweed, is an ideal biomass crop. Grown, on arid land, non-food crop, and easy to fertilize with natural waste water, duckweeds, offer the world an alternative to fossil fuels.

Market forces, never allow an opportunity to go unrecognized. China, for example, is an excellent candidate for rapid Duckweed utilization. Urban centers, in China, have an explosive problem in organic waste disposal. Mountains, of garbage, are being discarded, and cause an environmental hazard, attracting pests, and allowing disease bering pathogens to flourish.

Anaerobic digesters, are the best way to take organic materials, and "decompose" them to produce Methane Gas, and in-organic fertilizers. In-organic compounds, are those best available for plants to "uptake" for nutrients.

The "effluent" from the digesters is perfect fertilizer. This can be dried onsite, to remove moisture, using the "waste" heat from the digester. Methane gas, is produced by the anaerobic digester, and can be "burned" in a slightly modified engine for electricity production. This produces "waste" heat which can also be utilized for drying.

Organic wastes, are an issue of pollution, and lost resources. Two birds, with one stone, is to convert those organic wastes, before they are discharged into the environment, into Methane gas, for electricity production, and in-organic effluent, ideal for crop fertilization.

The world can make a transition from "ancient" Carbon, to atmospheric carbon. As market opportunities form, from higher and higher prices for traditional fuels, Duckweed, as an organic engine will gain market share.

Toxicity, bioaccumulation, and lack of access for all people, make a fossil-fuel powered civilization untenable. Using photosynthesis, as the Earth does naturally, as a mechanism to produce feed, fuel, and fertilizer is a path of equity, non-toxicity, and sustainability.

Duckweeds, have a major role to play in this industrial evolution from un sustainable practices, to sustainable access for all peoples, in all countries, using biotechnology, which makes the use of fossil-fuels, unattractive, unprofitable in comparison, and will be rendered obsolete.

The Duckweed revolution will not simply compete with Fossil fuels, it will obsolete fossil fuels, as the source for sustainable Carbon-based fuels, feeds, and fertilizers.

Chapter Nine - Duckweed Markets in Proteins, Carbohydrates, and Lipids (Oils)

Major markets, in base commodities, are valuable drawing billions of dollars per day, in trading. Proteins, vital for animal, and human feed markets are not only vital, but valuable in currently traded commodities.

Duckweed, has an active market in specialty nutraceutical market. Ideal, as a pet food for exotic pets, the "Wet" Duckweed product fetches $1-2 per ounce. At $2 per ounce, that's $32 per pound. A ton, of Duckweed, in this market is worth, Retail, about $64,000. Not bad, for a crop which grows from organic waste streams.

The value of Duckweed, on top of bioremediation of waste-streams, is Duckweed Biomass. Duckweeds, can be grown to maximize a particular group of desired molecules. If you value proteins, you can influence growth to favor protein production (less stress). Or, Duckweeds, can be grown under "stress" protocols, which favor Carbohydrate production. Trace amounts of anti-oxidants, and complete amino-acids, make the general "press-cake" of Duckweed useful in many products.

Specialty feed markets, such as for turtles, exotic birds, lizards, and amphibians, all benefit from the consumption of Duckweeds. High-value returns are possible, by simply growing Duckweed, (wet), and making available to the specialty feed markets.

Aquaculture markets, are expanding dramatically. Often, fish-farmers, use Soybean based products to supplement protein into the fish diet. Unfortunately, fish aren't evolved to digest Soybeans well, (too much fiber), and cause higher than average mortality rates in fingerlings.

Duckweeds, are perfectly suitable for fish consumption, and offer a complete amino-acid profile, and source for proper development of organs, flesh, and nervous system function. Once, duckweeds, are separated into proteins, carbohydrates, and lipids, each product stream can feed independent markets.

Proteins, for fish, poultry, swine, and cattle feed markets. Carbohydrates, for Ethanol production. And, omega III, and omega VI lipids for nutraceutical markets.

Duckweeds, grow rapidly, take up nutrients from waste-streams, and produce valuable amino-acids, proteins, carbohydrates, and lipids. The three principle markets, for Duckweeds, are Fuels, Feeds, and Fertilizers.

Duckweeds, represent an "industrial" revolution, in which waste-streams, are converted into Revenue-streams. Fast growing, diversified, and practical, Duckweeds, are the most valuable biomass crop in practical use. Long history, detailed studies, and centuries of experience, demonstrate Duckweeds, to be the next hot biomass crop.

A miracle of nature, Duckweeds, offer true commercial power to diverse economies, by decentralizing organic Feed, Fuel, and Fertilizer production.

Turning Waste-Streams, into Revenue-Streams, Duckweeds, will emerge as a superior biofuels, biofertilizers, and Nutraceutical source of primary production, which in maturity, will obsolete Fossil-Fuel based Fuels, Feeds, and Fertilizers.